SpringerBriefs in Earth Sciences

More information about this series at http://www.springer.com/series/8897

Anders Omstedt

Connecting Analytical Thinking and Intuition

And the Nights Abound with Inspiration

Photographs by Hillevi Nagel

Springer

Anders Omstedt
Department of Marine Science:
Oceanography
University of Gothenburg
Gothenburg
Sweden

ISSN 2191-5369 ISSN 2191-5377 (electronic)
SpringerBriefs in Earth Sciences
ISBN 978-3-319-27533-8 ISBN 978-3-319-27534-5 (eBook)
DOI 10.1007/978-3-319-27534-5

Library of Congress Control Number: 2015959942

Printed on acid-free paper

This Springer imprint is published by SpringerNature
The registered company is Springer International Publishing AG Switzerland

The Heart is the Capital of the Mind
The Mind is a single State
The Heart and the Mind together make
A single Continent

One—is the Population
Numerous enough
This ecstatic Nation
Seek—it is Yourself.

—Emily Dickinson

Preface

We are living in a world of increasing complexity in which accurately perceiving reality is increasingly important and difficult. Society's need to address "global grand challenges" requires that the scientific community initiate inter-disciplinary research often very unlike current education programmes, which are usually based on intra-disciplinary science. Improving our understanding of complex problems and communicating this understanding to a large group of people of differing backgrounds and educational levels, such as scientists from various disciplines, politicians, experts, students, and laypeople, represents a great challenge.

Analytical thinking is a powerful way of solving problems and has been used both to develop and destroy society and nature. However, we are only partly aware of the sum of our knowledge, and we are not restricted to analytical thinking. Art and our dreams are full of feelings and inspirations that can open up new dimensions in our lives. The great need for perceptual accuracy relies on our ability to connect analytical thinking with the intuition often manifested in dreams, literature, and art. This ability to interconnect these human capabilities, however, is not yet recognized in most educational programmes. The need to develop pedagogical methods for improving analytical, empathetic, and self-reflective thinking is stronger now than ever.

In this book, I investigate my own search for a new way to understand myself and my research area based on personal experience and on my work as a marine scientist in the field of oceanography and climate research. In recent decades, the marine environment has experienced serious damage and the scientific community has tried to act as a whistle-blower, though with few noticeable results. Society is entering a new era of conflict marked by increasing pressure on our natural resources and needs for new technology. It is imperative that we change our mentality, to become more environmentally aware, friendly, and caring. The existing trend towards ever greater knowledge fragmentation and increased competition requires that we step back and investigate our behaviour and driving forces in a more mindful way. At universities, the freedom to generate new knowledge needs stronger support, as does the understanding that the teacher's main missions are to generate new knowledge, educate, and to support the joy of learning.

This book illustrates how analytical thinking and intuition can be trained, and I propose a method for connecting both aspects in a systematic way. The dream group process developed by Montague Ullman, which includes training in analytical, empathetic, and self-reflection skills, is an excellent tool that should be included in university curricula, particularly for science students. In science, we learn to observe nature in an objective way, organizing measurement platforms and building mathematical models. Now is the time to employ methods that help us explore our driving forces, to improve our communication skills, and better understand how humans are influencing society and nature.

Acknowledgements

Over the years, many colleagues and students have inspired me to deepen my knowledge of the sea, and many talented dream group leaders and members have passed on their knowledge of dream work and literature. I would particularly like to thank Stina Hammar for introducing me to a new way of working with dreams and literature, to Marianne Blom and the “Pedagogisk Forum” for holding excellent and productive meetings, to Stig Östman for suggesting that I write a book on science and the humanities, to Elisabeth Bratt-Neuberg for her unflagging support and for suggesting that the book should be written in English, to Stina Nielsen and Kristina Elmén for many productive discussions, to the International Baltic Earth-BALTEX secretariat, particularly Silke Köppen, for their strong support, to the Baltic Earth-BALTEX scientific community, and to Stephen Sanborn at Proper English for editing the text.

Of particular importance to my life is my family, for giving me so much love and joy—thanks Britt, Herman, Mikael, and Ragnhild for all you have given me.

Acknowledgments

Contents

About the Author

Anders Omstedt is a professor at the University of Gothenburg, Department of Marine Sciences: Oceanography. His main research interests are the ocean climate and ice-covered water bodies, particularly in coastal seas and lakes. He was a part-time scientific coordinator at the Swedish Institute for the Marine Environment for five years and previously occupied various positions at the Swedish Meteorological and Hydrological Institute.

He has long experience working with dreams and is a dream group leader approved by the Swedish Dream Group Forum.

About the Photographer

Hillevi Nagel is a photographer educated at the University of Gothenburg, Valand Academy, and Biskops-Arnö Adult Education School, where she studied documentary photography. She has worked on several book and exhibition projects featuring pure and simple narrative photos. She has long experience in working with dreams, mainly following Ullman's dream group method.

Abstract

In this book, analytical thinking and intuition are illustrated based on my personal experience of working in marine research and in dream groups. The book will stimulate students and scientists to improve their thinking and communicating skills and to increase their joy in learning more about being human. This guide on connecting analytical thinking and intuition is inspired by the dream group method developed by Montague Ullman, in which a group of six to eight people systematically and carefully helps the dreamer to appreciate his or her dreams. The importance of the good memories that can be trained through recalling one's dreams is discussed in relation to science and literature. The analytical scientific process and intuition are illustrated in several figures and photos, and creative and stimulating exercises are suggested.

Keywords Montague Ullman · Earth sciences · Oceanography · Dream sharing groups · Communication science · Memory training · Analytical thinking · Creativity · Intuition

Chapter 1
Introduction

Most people know that when struggling with a problem, the answer may come more easily after a good night's sleep. While we are sleeping, our minds continue to work and organize the impressions gathered in daytime. It is easy to find examples of brilliant scientists who made discoveries inspired by dreams. For example, Dmitri Mendelev was able to organize the periodic table of the chemical elements based on their atomic numbers after hard work and a dream. In daily life as well, a good night's sleep can help us solve many problems. Many of us have learned that creativity is associated with hard work. This is also what we teach our students when loading them with more and more work during the education process. It has surprised me that we, scientists and teachers, so rarely discuss the creative process and how to explore it. Even today, this topic is often not even included in university curricula. Most people are interested in the creative process and have varied idiosyncratic views of how it can be improved, but there is little systematic discussion of how to develop methods to promote creativity. Instead, we feel driven to explore, in an unplanned, unsystematic, un-self-aware way, various inspiration initiatives or to apply new pedagogical methods without deepening our own knowledge of creativity.

One aim of this book is to illustrate how analytical thinking and intuition can be trained and connected by observing how dreams play a role in consolidating information in our brains, giving us a systematic method of supporting our own creativity. Another aim is to illustrate how the beauty and power of our subconscious mind supports our conscious mind. The presentation will follow the various steps of dream group process inspired by the method of Ullman (1996), Siivola (2011). The book will not review the practical aspects of working with dreams, as several useful books on the subject are already available. Instead, the reader is referred to the references in each chapter as a guide to further reading, while a short summary of Montague Ullman's dream group process is given in Appendix A. In Appendix B, the interested reader can attempt some exercises that illustrate how analytical thinking and intuition can be connected, while Appendix C provides some ideas for solving these exercises.

A. Omstedt, *Connecting Analytical Thinking and Intuition*,
SpringerBriefs in Earth Sciences, DOI 10.1007/978-3-319-27534-5_1

Fig. 1.1 The beauty of our natural resources is illustrated by a satellite image from 1 April 2004 (from the SeaWiFS NASA/Goddard Space Flight Centre in the USA, http://visibleearth.nasa.gov/)

In this book, I will explore dreams by citing a number of examples from my own experience. At the same time, I will connect the discussion with the scientific process as I have experienced it in my work as a marine scientist. All direct references to dreams are presented in italics. Working scientifically can be regarded as exploring the unknown using our conscious minds, for example, in understanding how our coastal seas function (Fig. 1.1). Systematic scientific work requires many years of education and an interest in asking new questions. Working with dreams is like singing in a new language. This language differs from that of analytical thinking and expresses feelings, often in a striking and emphatic way. Most of us have the ability to sing and are fascinated by the beauty of singing. Dreaming has similar values and does not belong to any special expert group. Instead, it is a resource that is freely available for all to explore and a gateway to a deeper understanding of ourselves, humanity, and creativity.

References

Siivola, M. (2011). *Understanding dreams: The gateway to dreams without dream interpretation.* New York, NY: Cosimo Books.

Ullman, M. (1996). *Appreciating dreams: A group approach.* Thousand Oaks, CA: Sage Publications.

Chapter 2
Catch the Dream

When I was around thirty-years old, I felt a strong need to develop a new language that could better express my feelings and ideas. I felt frozen and trapped by many expectations. The Swedish poet Stig Dagerman helped me put words to my feelings in his 1953 poem "Ice age":

You have no soul when it's cold,
You have an icy little business stamp.
You might almost wish you could melt
and become a river, like the Nile for example.

But here no one becomes what he wants to be.
You stand like a statue in the snowy landscape.
Suddenly one day you have frozen solid
and every time you open your mouth you spout ice.

And then everything happens as foretold.
In front of some fire your icy profile melts.
The hostess brings nothing but a rag.
You become a puddle, though you wanted to be the Nile.

The poem was written before the Aswan Dam was built to regulate the Nile River, the world's longest river. Before the dam was built, the Nile River flooded every year, bringing water, nutrients, minerals, and silt to fertilize and build up the delta, making the Nile Valley and Delta ideal for farming. Today, a combination of regional anthropogenic changes, such as the Aswan Dam and climate change induced by increased greenhouse gas levels, is seriously threatening the Egyptian coast. Later in my life, I had the opportunity to work on climate change in the Mediterranean Sea area with Mohamed Shaltout from the University of Alexandria. We demonstrated that even a slight sea-level rise could greatly damage the Nile Delta region (Shaltout and Omstedt 2015). This led to a call to organize, with the involvement of the wider society outside the scientific community, effective integrated coastal zone management to prevent future flooding. Such interaction is often needed in the Earth

A. Omstedt, *Connecting Analytical Thinking and Intuition*,
SpringerBriefs in Earth Sciences, DOI 10.1007/978-3-319-27534-5_2

sciences, creating a need for new communication skills not touched on in most science curricula but discussed widely in relation to climate change. In my early thirties, I wrote a short article in our union newsletter, accompanied by Stig Dagerman's "Ice Age", expressing my frustration at being trapped in an old mode of work organization. My search for a new language had started.

I had defended my Ph.D. thesis about cooling and ice formation in the ocean while working at the Swedish Meteorological and Hydrological Institute, developing forecasting models for the ice-breaking service. The scientific work was a great pleasure: in particular, testing theories about the sea against field observations evoked unfamiliar and exciting feelings. The possibility of integrating mathematical expression into computer codes and of testing theories directly against observations inspired me greatly. However, I did not feel satisfied or free and, needing to redirect my life, attended a summer course to work on dreams, myths, and literature. The first night I dreamt that *I was standing outside a department store and inside I could see lots of interesting boxes. When I went to the main entrance, the door was locked.* The night after this dream, I dreamt that *I was bitten by a scorpion on my Achilles tendon.* My interest in dreams had awakened, and I realized that the key to my dreams was my own vulnerability and feelings. That first summer course was held at Biskops-Arnö Adult Education School, beautifully situated on Lake Mälaren outside Stockholm, by Stina Hammar, who has written several books about literature and dreams. From that beginning, attending summer courses at various places around Sweden and working in dream groups gave me an education in dream appreciation. At the same time, I was exploring my academic field and deepening my understanding of how the seas function.

Reference

Shaltout, M., & Omstedt, A. (2015). Sea-level change and projected future flooding along the Egyptian Mediterranean coast. *Oceanologia, 57*(4), 293–307.

Chapter 3
Finding Words for Feelings

I am out sailing in a large modern sailing boat. We have strong winds behind us and all the sails are unfurled. I am on deck tending to the sails. Suddenly, I see my young son climbing up onto the deck and then falling into the sea. I rush to the tiller and turn around the sailing boat. A man helps me to bring my son on board.

Dreams are a mode of internal communication that can, through their feelings and symbols, transfer knowledge from our unconscious to our conscious domains. Putting words to one's feelings is of fundamental importance, as feelings are involved in all kinds of communication between these two parts of our psyche. The dream described above evoked feelings of happiness, wildness, stress, fear, alarm, and relief. At the same time, the dream was full of symbols related to boats, sails, steering, sons, and father. Symbols have no standard interpretations but instead are replete with allusions and alternative meanings. Sailing with all sails unfurled could indicate that all the dreamer's capacities are being used and that life is rushing past. A son falling into the sea may indicate that the dreamer is in danger of missing something important, while rescuing the son could mean that, helped by another inner ability, the dreamer was able to save something important. In general, the dreamer is the one who can best understand his or her dream, though the dreamer often needs help from a dream group to link the dream's emotive and symbolic content to the dreamer's situation in life. The emotional context of the dream may then guide the dreamer to a deeper understanding of the dream's symbols and emotive content.

When I had this dream, I had just started a project together with a friend and a small group. The first meeting was a catastrophe as the group was headed in two different pedagogical directions. After my dream, I realized that I had to speak with my friend and take over project leadership. This project was of great importance to me and the class could not be led in two directions. My friend admitted that she was very tired and was in fact pleased that I wanted to lead the project.

A. Omstedt, *Connecting Analytical Thinking and Intuition*,
SpringerBriefs in Earth Sciences, DOI 10.1007/978-3-319-27534-5_3

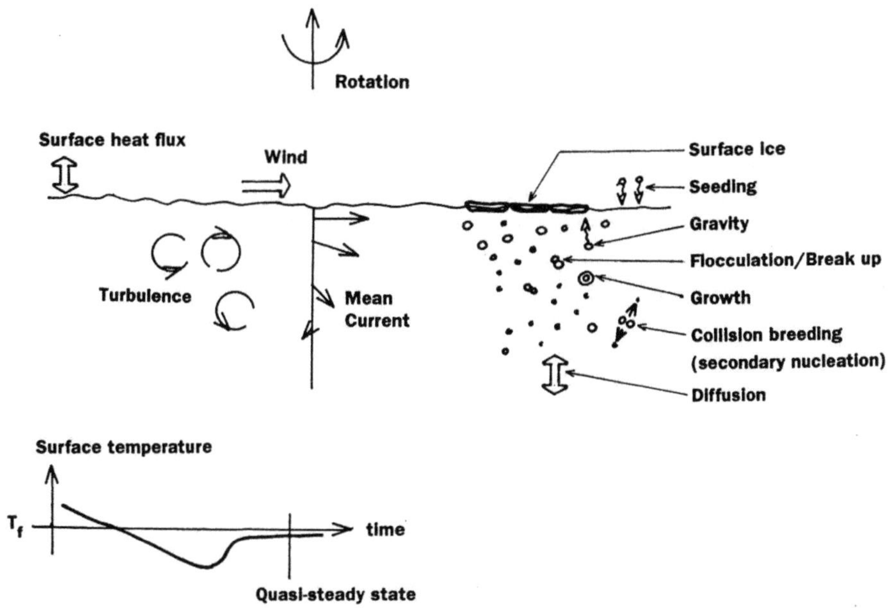

Fig. 3.1 Science often starts by making a simplified picture illustrating certain important factors of a problem (from Svensson and Omstedt 1998)

Dreams come to us as a series of images imbued with feelings and symbolic value. In this dream, the images guided me to better understand my situation and to improve my teaching and leadership.

At the time of this dream, I had published a number of scientific papers that had contributed to my understanding of sea ice; I had learned how sensitive sea ice conditions were to any changes in climatic conditions (Omstedt and Nyberg 1996). The scientific community was exploring many interesting physical processes, such as how ice forms in turbulent water (Svensson and Omstedt 1998, Fig. 3.1) and how the ocean transfers heat to ice (Omstedt and Wettlaufer 1992). Another major question was how to model sea ice and sea ice ridging. I was happy to be part of the research into these issues and to apply new knowledge in Baltic Sea ice models. The icebreaking service was following the discussion about climate change but, even though the scientific community had already started to consider global warming, the service stuck with using historical statistics as a tool for determining the size of the icebreaker fleet. Very few were interested in Baltic Sea ice and regional climate change consequences in the 1990s, but interest from the scientific community was increasing and ice modelling had prompted me to think further.

References

Omstedt, A., & Nyberg, L. (1996). Response of Baltic Sea ice to seasonal, interannual forcing and to climate change. *Tellus A, 48*(5), 644–662.

Omstedt, A., & Wettlaufer, J. S. (1992). Ice growth and oceanic heat flux: models and measurements. *Journal of Geophysical Research, 97*(C6), 9383–9390.

Svensson, U., & Omstedt, A. (1998). Numerical simulation of frazil ice dynamics in the upper layer of the ocean. *Cold Regions Science and Technology, 28*(1), 29–44.

Chapter 4
Meaning of Symbols

I am steering a motorboat from the Baltic Sea into the Kattegat. Several of us are on deck. Suddenly, I see a military ship coming towards us. It approaches us and a Nazi officer jumps on board. Our boat has been seized and, inside the cabin, the crew starts to organize resistance. The children start an orchestra using pots and kitchenware, and sing along with their playing. The Nazi officer becomes afraid, leaves our boat, and we are free. I resume steering the boat and carefully watch the sea.

Dream symbols can illuminate feelings in both dramatic and comic ways. I had the above dream at about the turn of this century, just after I moved from one job in Norrköping, a city near the Baltic Sea coast, to another job in Göteborg, a city on the Kattegat coast. The motor boat symbolizes my situation in life, travelling from one city to another. The symbol of the Nazi military ship seizing my life illustrates the pressure I was experiencing in taking this new job. The way I defended my personality is illustrated by the children's actions, symbolizing a different and more creative approach to life. The dream bolstered my confidence in my new work but also told me that I had to be careful and watch out for "occupation" by problematic feelings. The dream illustrated two attitudes inside my mind, a controlled and a playful one, and the importance of understanding the difference between them.

It was not until 2003 that we in the marine science community developed a good understanding of the physics of our coastal seas (Winsor et al. 2001; Omstedt and Axell 2003). Since that time, we have had the tools to address how water temperature, sea ice, and salinity may have changed in the past, how they are behaving in the present, and how they may respond to possible future climatic conditions. The idea that captured my interest was that to understand a century forward in time, we must understand at least ten centuries back in time. I therefore started to guide part of my research group into historical studies, and for several years we were able to find lots of interesting information about how the seas have changed (Eriksson et al. 2007; Hansson and Omstedt 2008). This scientific playground was creative and satisfying and nobody asked us for results; instead, we were free to go our own way.

A. Omstedt, *Connecting Analytical Thinking and Intuition*,
SpringerBriefs in Earth Sciences, DOI 10.1007/978-3-319-27534-5_4

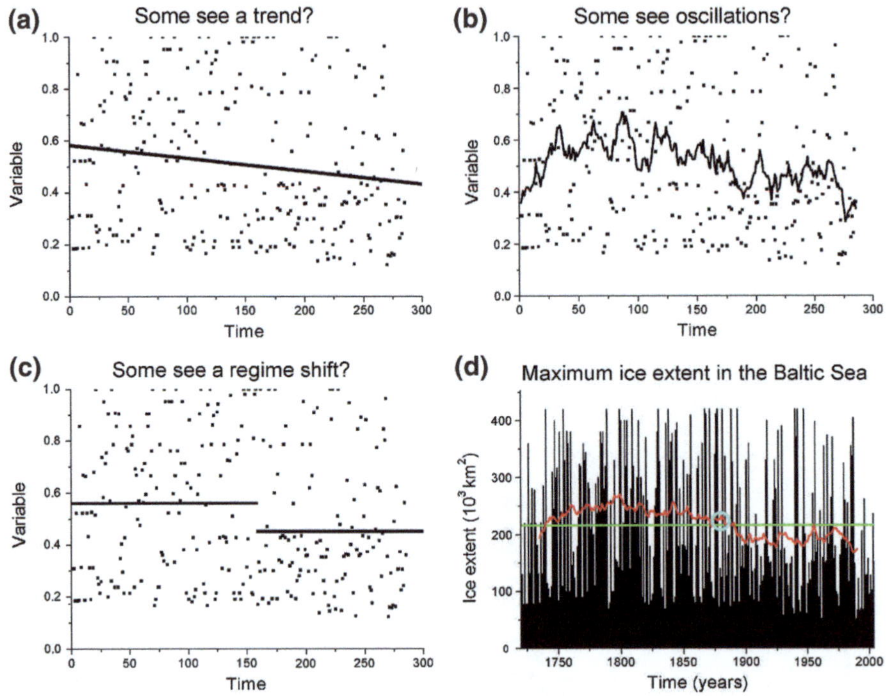

Fig. 4.1 Observations can be analyzed in terms of trends, oscillations, jumps, or regime shifts. Scientists often prefer to present the single interpretation best supported by wider society's discussion of the topic. In this figure, the same data are used and normalized in panels **a–c**. The original dataset is presented in **d** (redrawn from BACC Author Team I 2008)

At the same time, the larger scientific community was becoming polarized into groups convinced that modern climate change was outside historical norms and was attributable to fossil fuel burning and groups that were claiming that the data and models still harboured severe biases. The science was strongly influenced by funding, political considerations, and the mass media. In the 1990s, many scientists liked to present data and focus on oscillation aspects; by the 2000s, many started to discern trends and later regime shifts in their data (Fig. 4.1). The scientific community was not independent of society, but instead showed itself to be closely connected to societal trends.

I realized that only careful historical studies could discern climatic variability on centennial time scales; so we delved into many different sources of information, ranging from instrument data to historical documents and paintings. Several important multi-proxy datasets were developed and assembled into freely available datasets. My research group learned that climatic variability was quite large on the regional scale and that even during what we call the Little Ice Age, several warm periods occurred. This left us suspecting that perhaps our present time was not unique in terms of temperature. In addition, long-term studies of river runoff to the Baltic Sea indicated climatic variability but no trend on the centennial time scale.

We later found a beautiful scientific report of research vessel observations from 1877 (discussed in Omstedt and Veiderpass 2015). This was the first modern expedition around the whole coast of Sweden, and interpreting these data in relation to current climatic conditions made it clear that only modest changes had taken place, deep water temperatures being only about one degree colder in 1877 than today and accompanied by no change in salinity. The temperature changes in our own sea had been quite modest, in sharp contrast to the general lay opinion. We could discern only a slight warming from the end of the nineteenth century to the present, and changes related to this warming were evident in factors such as snow and ice. In the first regional assessment it was stated: "The Baltic Sea region is warming, and the warming is almost certain to continue throughout the twenty-first century. It is plausible that the warming is at least partly related to anthropogenic factors. So far, and as is likely to be the case for the next few decades, the signal is limited to temperature and to directly related variables, such as ice conditions. Changes in the hydrological cycle are expected to become obvious in coming decades. The regional warming is almost certain to have a variety of effects on terrestrial and marine ecosystems, some being more predictable (such as the changes in phenology) than others". This assessment summarized the understanding of a large group of active scientists without any influence from other groups in society (BACC Author Team I 2008).

Being part of a large scientific community was inspiring; at the same time, my own understanding of the science was being tested through the challenges of communication and writing. In the assessment work, we were seeking all available knowledge in the literature, not the "best" knowledge, and we were working with both scientific consensus and scientists' varied individual opinions. For me this was a democratic and healing effort, giving me trust in the scientific process.

References

BACC Author Team I. (2008). *Assessment of Climate Change for the Baltic Sea Basin.*, (Series: Regional Climate Studies) Berlin: Springer.

Eriksson, C., Omstedt, A., Overland, J. E., Percival, D. B., & Mofjeld, H. O. (2007). Characterizing the European sub-arctic winter climate since 1500 using ice, temperature and circulation time series. *Journal of Climate, 20*(21), 5316–5334.

Hansson, D., & Omstedt, A. (2008). Modelling the Baltic Sea ocean climate on centennial time scale: Temperature and sea ice. *Climate Dynamics, 30*(7), 763–778.

Omstedt, A., & Axell, L. (2003). Modeling the variations of salinity and temperature in the large gulfs of the Baltic Sea. *Continental Shelf Research, 23*(3–4), 265–294.

Omstedt, A., & Veiderpass, V. (2015). Den hydrografiska resan 1877 [The hydrographic expedition 1877]. In H. Svedäng (Ed.), *HAVET 1888 [The sea 1888]* (pp. 19–21). Gothenburg, Sweden: Swedish Institute for the Marine Environment.

Winsor, P., Rodhe, J., & Omstedt, A. (2001). Baltic Sea ocean climate: An analysis of 100 year of hydrographic data with focus on the freshwater budget. *Climate Research, 18*(1–2), 5–15.

Chapter 5
Triggers

Dreams start from feelings generated in our daily lives but do not end there. Certain feelings experienced during the day remain with us and enter into the domain of sleep. These lingering feelings play an important role in dream creation and often obtain energy from important and still-unresolved personal issues. All of us bear the burden of earlier life experiences that make us vulnerable, at the same time as we are facing a future replete with challenges. We often need to suppress feelings of, for example, stress and anger to be able to cope with daily life. This internal stress often triggers dreams expressed in a language of feelings.

My first contact with systematic dream work was at the summer course at Biskops-Arnö mentioned in Chap. 1. At that time, I did not know much about dreams and I was not prepared for dream work when I began the course. The summer course focused on literature, dreams, and myths and the only thing I was prepared for was the literature part. The first day, I met a group of new people and, with an experienced dream group leader, started to work on dreams. In the group, a woman presented her dream and, during the session, we worked through it. The dream group work was very creative and the dreamer was open to the process and to its potential to alter her understanding of her situation in life. I was greatly impressed. This feeling stayed with me, and in the evening when I went to bed I dreamt about *a department store that contained lots of interesting boxes but whose front door was locked*. My experience that day, of lacking knowledge of dreams but being impressed by their strength, was illustrated by the locked department store containing many potential gifts. In addition, my feelings and experience before the dream may have been amplified by my earlier stress of being an outsider to the group and ignorant of dream work. These feelings triggered the dream and motivated me to explore my understanding of dreams in greater depth. These feelings were not that different from what one can experience as a scientist when mainstream developments are attracting most interest and one decides to go on one's own independent way.

A. Omstedt, *Connecting Analytical Thinking and Intuition*,
SpringerBriefs in Earth Sciences, DOI 10.1007/978-3-319-27534-5_5

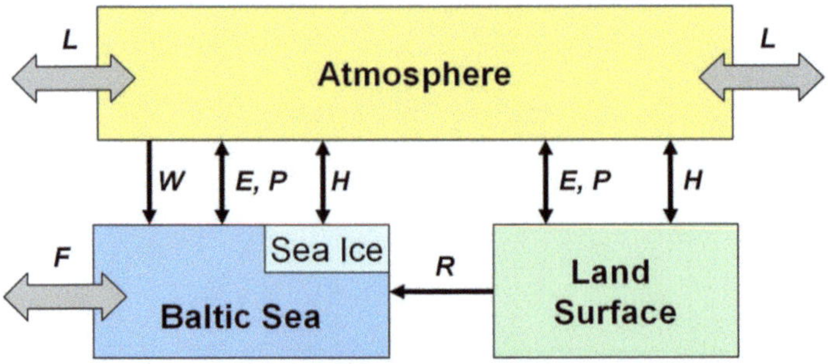

Fig. 5.1 This simplified model of nature was the trigger for ten years of research within BALTEX (Raschke et al., 2001), a joint international programme of Baltic Sea research focusing on physical processes. Here, *W* denotes wind, *H* heat flux, *E* evaporation, *P* precipitation, *F* inflows and outflows, and *R* river runoff (figure courtesy of Marcus Reckermann)

In my own scientific field, oceanography, understanding starts with simplification (Fig. 5.1) and scientists who have successfully applied this method, such as Henry Stommel, who explained important aspects of Gulf Stream dynamics, were our heroes. On the other hand, many other scientists, particularly from the marine biology community, glorified concrete details. When communicating to people that focusing solely on the most important aspect could mean the overlooking of details? The novel *Siddhartha* by Herman Hesse tells how a young man started to deepen his understanding by leaving behind a privileged life to discover his own ideas. Siddhartha wanted to discover the roots of his own feelings, believing that with this knowledge he could understand the causes of phenomena. He recognized early on that what he knew least about was himself. From that insight, he embarked on a long journey to know himself. Afterwards, Siddhartha formulated the maxim that for every true idea there is an opposite one that is also true. Maybe this also applies to both science and dream interpretation: for every really good simplification there is an opposite complex solution that is also true—an exciting possibility that triggered my curiosity.

Reference

Raschke E., Meywerk J., Warrach K., Andrea U., Bergström S., Beyrich F., Bosveld F., Bumke K., Fortelius C., Graham L.P., Gryning S.-E., Halldin S., Hasse L., Heikinheimo M., Isemer H.J., Jacob, D., Jauja I., Karlsson K.-G., Keevallik S., Koistinen J., Van Lammeren A., Lass U., Launianen J., Lehmann A., Liljebladh B., Lobemeyr, M.,Matthäus W., Mengelkemp T., Michelson D.B., Napiorkowski J., Omstedt A., Piechura J., Rockel B., Rubel F., Ruprecht E., Smedman A.-S. & A., Stigebrandt. (2001). The Baltic Sea Experiment (BALTEX): A European Contribution to the investigation of the Energy and Water Cycle over a Large Drainage Basin. *Bullertin of the American Society*, *82*(11), 2389–2413.

Chapter 6
To Be Touched

Dreams, poetry, art, literature, music, religion, and science all engage different parts of the human creative faculty and we can be touched by various aspects of these endeavours. Perhaps one can say that the joy of learning is one of the most important drivers of progress in life. At the same time, our minds change and sometimes open us up to human kindness. I was first touched by dreams when I dreamt that a scorpion bit my Achilles tendon. Since then, I have studied my own and others' dreams for almost three decades. Parallel to this, I have worked as an oceanographer exploring how the sea functions. The strong connections between my scientific work and my dream work have surprised me. Starting with a Ph.D. thesis on cooling and ice formation in the sea, I was struck by a desire to understand how ice interacts with sub-polar and polar seas. When modelling ice formation in turbulent water due to the formation of frazil ice (i.e. small ice crystals growing in water), my curiosity was first engaged by how ice crystals grow and connect to each other, forming surface ice. Then understanding how ice drifts, how ice forms ridges and leads, how ice grows and melts, and how this knowledge could be transferred to other people using forecasting methods kept me busy for a decade.

At the same time, I started recording my own dreams and studying others' dreams in literature and dream groups. The theme of several summer courses I took was meeting oneself through other people, and the methods for doing this were Ullman's dream group method, literature study, and dancing. Both the scientific and dream work processes have given me methods for exploring external seas and inner feelings. Through inspiration from a Canadian marine science colleague, I formulated my rules for conducting science, which I later taught to a number of students. To succeed in science one must consider three things that can be summarized as follows: push (but gently), focus, and do not hurt yourself. These three rules gave me a strong basis for scientific work. A similar strong method in dream work is when the group takes a dream as its own and starts to explore the subject's feelings as conveyed in the dream and its symbols. By relating the problem or dream to one's own experience and by investigating and understanding one's feelings, we systematically learn how

A. Omstedt, *Connecting Analytical Thinking and Intuition*,
SpringerBriefs in Earth Sciences, DOI 10.1007/978-3-319-27534-5_6

to work with both sides of our minds, the analytical and intuitive parts. We also increase our capacity for empathetic understanding by listening to the dreamer and relating his or her experience to our own. This knowledge is increasingly important in science as well as in communicating with experts of different backgrounds and with the general public, all of whom communicate in different ways.

Growing challenges for scientists are to understand how humans interact with nature, staying open-minded, and actively listening to society's need for change without becoming soul-hardened. I have worked with many scientists and students on the functioning of the ocean climate system, conducting studies of reconstructed past, present, and projected future climatic conditions and the factors controlling them. My scientific approach was based on a process understanding and on developing numerical models that successively add new processes (Omstedt 2015). I believed that by understanding important processes we could build up a systemic understanding of our environment. We learned that much was unknown and we were distressed by the fact that some were trying to "sell" an oversimplified view. We initially examined the natural physical system and later its coupling with chemistry, biology, and human interaction. Our modelling became increasingly complex and it took us over a decade to produce a scientific instrument that could systematically investigate human impact on the sea.

From this work, it was clear that we as humans strongly interact with our marine environment, and in a mostly negative manner (e.g. Reed 2015). The message of our scientific work could perhaps be summarized as follows: We seem to change our mentality when entering the seas. On land, we have a caring mentality, but when we leave behind the coast for the sea, we seem to switch to the mentality of pirates. Hidden in our minds are the old assumptions that the seas are infinite, that fish are free for the taking, it is better to grab everything now instead of waiting, and that no one is aware of what we are doing when we dispose of waste in the sea. Most people are now moving into cities and our coasts are divided between highly populated centres and—in Sweden, anyhow—summer resort villages that are lifeless in the winter. Our estrangement from the seas is increasing. We are increasingly becoming mere exploiters of marine natural resources and are forgetting that the coastal seas represent great resources for our overall development, not just physical resources for food, shipping, and recreation but also imaginative resources supplying inspiration, instilling respect, and offering tranquillity. Current human management is failing to restrict exploitation and is unable to preserve the seas in a healthy state. Instead, overfishing, eutrophication, poor pollution control, plastic waste, shipping, climate change, and ocean acidification are severely threatening the marine environment.

Delving deep into ocean science has taught me that we must change our thinking and become more mindful and careful of our natural resources. We have entered a new era in which nature is greatly dependent on what we as humans are doing: nature can no longer recover by itself. Scientifically, we now have methods that can guide us towards better management; the limitations are now found in society and in our minds.

References

Omstedt, A. (2015). *Guide to Process Based Modelling of Lakes and Coastal Seas* (2nd ed.). Cham, Switzerland: Springer International Publishing.

Reed, C. (2015, 31 January). Dawn of the plasticene. *New Scientist, 3006*, 28–32.

Chapter 7
Exploring the Unconscious and the Unknown

I am out sailing in an old sailing ship. We are sailing towards a large city and we are in a hurry – maybe someone is following us? We are under full sail, and suddenly the mainmast is breaking. On board are the male crew as well as women and children. The women and children must be rescued first. Three children are inside the ship, and two are weeping because they are surrounded by flying bats, some of which are perching on the children. The third child explains that, because of the light from above, the bats like to stay on the boat and that they are not dangerous.

The dream evokes feelings of freedom, stress, danger, and fear that give way to a feeling of relief after hearing that the bats are not a threat. The three children inside the old ship are the two life phases I contain within myself: the young child and the old parent. The bats are dreams and the light can be seen as the wisdom that dreams bring. A new side of me, represented by the third child, understands that dreams do not mean danger and tries to allay the fear felt by the other two children. I need to tell these two about the positive side of dreams. Often in dream groups, I have experienced how strongly dreams support us and have understood that they are not dangerous, even though they can be violent. By putting words to dreamed feelings and experiences, we can improve our mental health and develop more accurate perceptions. Inexperienced dreamers often interpret their dreams in a negative light, while dreams in fact often contain strongly supportive messages that can help dreamers cope with their lives in new ways. A supportive dream group can carefully emphasize the positive messages in dreams. The old ship and the breaking mast symbolize that part of me that needed to find new ways to revitalize. Reconnecting to my younger self supported me in my inner journey and challenged me to communicate my experiences.

I am out walking towards a market where I see a friend and another man in discussion. I walk towards them and find that my friend has a baby trolley containing a boy sitting in what looks like a large egg. I turn towards the child and take him out of the trolley. We start speaking with each other and I ask him how he is? The boy replies that he is happy, and I feel that it is very important to communicate with him. He looks ragged and in need of care.

A. Omstedt, *Connecting Analytical Thinking and Intuition*,
SpringerBriefs in Earth Sciences, DOI 10.1007/978-3-319-27534-5_7

The dream awakens feelings of caring and involvement. The egg-shaped baby trolley symbolizes the birth of something new and the boy is a part of me that I have not yet nurtured. I am now more open to my own inner experience.

It is early summer and we—my family and I—are out walking. A young boy is following us and I feel that I need to take care of him. We come to a café in the park and sit down around a table. We are speaking together and I notice that the boy has picked up a flying toy, which he throws into the air. It is a beautiful dragonfly that flies around our table and we are all excited, although I know that it is just a toy. The boy then picks up a metal bullet and throws it towards me. I am afraid, but nevertheless catch it out of the air.

The feelings evoked by this dream are joy at being with my family, curiosity at the entrance of the unknown boy, a mixture of concentration and amazement at seeing the airborne dragonfly, withdrawal and perhaps disappointment when realizing that it is just a toy, and fear when the boy throws the bullet. My family symbolizes a well-known and important part in my life, while the boy is a part of me that is now calling for my attention. The attack from the boy indicates the importance of attending to my new, growing consciousness.

Two foreign children come to me bearing a letter and a cheque for 90,000 pounds. In the letter, the sender asks me to take care of these two children.

These dream images awaken feelings of surprise, dedication, and responsibility. The symbols of the foreign children and the letter represent two unknown parts of me that I need to address, while the cheque indicates that doing so will bring great rewards. The number 90,000 is large and at the same time is a phone number formerly used in Sweden to call for emergency services.

The scientific process concentrates on how to expand on currently available knowledge. It is thus a risk project with unknown results as well as a search for true answers. It cannot promise to provide society with solutions or products that are easy to apply. Instead, research leads to new questions at new levels of understanding. With reference to the seas, it is fundamental to know what controls physical processes such as currents, mixing, waves, temperature, salinity, and ice formation. One cannot understand what is changing in a marine system without understanding these factors, but this is not the whole story. Observations of oxygen indicate strong declining trends in many coastal seas, which calls for an understanding of how plankton generate oxygen through photosynthesis, how the air–sea flux mixes oxygen into the surface water, and how dead biological material reduces oxygen levels through mineralization. Nutrients play an important role in this, being used during plankton blooms and released during mineralization.

At the same time, the composition of gases in the atmosphere is changing due to human activity and concentrations of carbon dioxide are increasing. What effects these changes will have on our terrestrial and marine ecosystems are still mostly unknown and call for new research efforts (Fig. 7.1). The increased eutrophication in coastal seas and in the Baltic Sea is partly due to human activities on land, such as farming, land-use change, and urbanization, which send increasing nutrient loads to the sea. Attributing the causes of eutrophication has been debated by scientists for

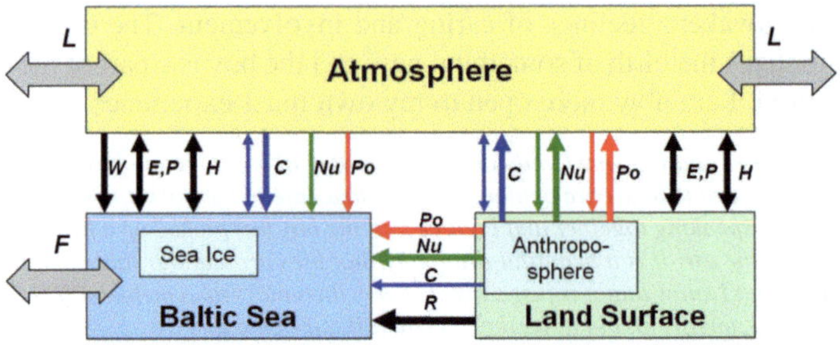

Fig. 7.1 As the research addressing both climate and environmental change became more complex, this simplified view was the trigger for another ten years of research (Reckermann et al., 2011). Here, *L* denotes lateral exchange with the atmosphere outside the region, *W* wind stress, *E* evaporation rate, *P* precipitation rate, *H* heat and energy fluxes, *R* river runoff, *F* inflows and outflows through the entrance area, *C* carbon fluxes, *Nu* nutrient fluxes, and *Po* pollutant fluxes (figure courtesy of Marcus Reckermann)

several decades, and research groups have been in conflict for years as to who is right. At the same time, science has become much more international; for example, in the Baltic Sea region, with the collapse of the Soviet Union and the formation of the European Commission, great progress has been made in promoting a more open scientific attitude. Scientists are not different in kind from other people and, like most people, seek personal and group advantages. The scientific process, however, includes many self-regulating mechanisms to correct flawed information. The feature most characteristic of valid science is reproducibility. If scientists from different research groups cannot reproduce new results, they must conclude that they are invalid. Open and respectful communication and free information flows are therefore fundamental to any efforts to explore the unconscious and the unknown. Here the society needs strong improvements.

Reference

Reckermann, M., Langner, J., Omstedt, A., von Storch, H., Keevallik, S., Schneider, B., Arheimer, B., Meier, H.E.M., and B., Hunicke (2011). BALTEX–an interdisciplinary research network for the Baltic Sea region. *Environmental Research Letters* 6(2011) 045205 (11pp). DOI: 10.1088/1748-9326/6/4/045205.

Chapter 8
Inspiration

I am on the Swedish west coast in an older time. We are staying in a house by the sea, waiting for some people coming from the sea. I ask my brother to prepare for the visit. I see now that the guests are coming—an old sailing ship with two masts is approaching. When the ship comes closer the water level sinks and a large stone becomes visible. The crew sails the ship near the stone. They throw an electric net over the stone and blast it into thousands of pieces that, like a white cloud, disappear in the air. I tell everyone in the house to stay away.

Many feelings are aroused: anticipation when waiting for the ship, peace and contentment because of the natural environment, happiness at seeing the ship, being impressed at the crew's skill, worry when the water level sinks, curiosity when the net is brought out, relief when the stone is blasted, and caution at the end of the dream.

The old house on the Swedish west coast is one of our marine research stations, the Bornö Marine Station. It was built over a hundred years ago and has been an important meeting place for marine scientists, in particular oceanographers. The old ship and historical setting refer to old methods. The stone represents barriers to sound marine management promoting healthier sea conditions. The electric net symbolizes modern means of eliminating barriers to communication, means such as the Internet. The cloud indicates that I need to stay away and wait for clear air.

This dream occurred when I was the scientific coordinator of the Swedish Institute for the Marine Environment (SIME) and was launching a new Internet portal for marine information. Overall Swedish marine management was a mess, with strongly competing actors working at various national institutions and universities. SIME leadership was disorganized and harboured various hidden agendas. The stone symbolizes the barriers in the main players' minds, but the dream's inspiration suggests that the Internet will blast exclusivity and local competition away, opening up better lines of communication. This dream fostered two attitudes within me, an active one, challenging barriers by using the Internet, and a more careful one,

A. Omstedt, *Connecting Analytical Thinking and Intuition*,
SpringerBriefs in Earth Sciences, DOI 10.1007/978-3-319-27534-5_8

focusing on avoiding harm. Soon after this dream, I left SIME to concentrate more on science-related issues and writing.

I see a woman on a horse. She bends down and licks away blood from the horse. I am up in the air looking at the horse and see three American Indians coming towards the horse. I understand that this is what is needed to cure the horse.

The dream, which came to me just after I had left SIME, illustrates how my strength and inspiration, represented by the horse, had suffered. The woman symbolizes the need for compassion and healing. The dream also points towards a need for a new perspective and for inspiration from a new source of energy represented by the three American Indians. The night after this dream, another dream occurred:

I am up in the Alps and see lots of horses coming towards me. In the Alps I am supposed to meet a well-known Swedish climate researcher. He is not there. Instead, a well-known Swiss climate scientist meets me and guides me to a village and restaurant. Before eating we climb up on a rock where I meet C. G. Jung. He takes up a bottle containing crystal-clear water and drinks. Then he gives me the bottle.

Being high up indicates that I am searching for an alternative way, hoping for guidance from my scientific understanding represented by the two climate researchers who have been important persons in my scientific career. Instead, I meet another part of my mind represented by C.G. Jung, the Swiss psychotherapist well known for his deep understanding of dreams. Drinking crystal-clear water with him strongly encouraged me to continue exploring my inner space.

I am in an art gallery working with clay and paper, forming faces with various hair styles. I am helping a woman mould a boy's face and hair. I make a nice-looking fringe and when I am ready I start singing: "My love, you are like a rose a ...". I am singing in a dark, deep voice and I realize that the people surrounding me like my singing.

This dream, which came to me two weeks after the horse dream, reminds me of my love of life, the inspiration I take from it, and of my joy at being part of society. The dream connected me to my own source of inspiration and revived me. I felt freed from any feelings of bitterness.

I am at the seashore and running along a bridge with my daughter. I try to jump into the sea but instead stay in the air. My daughter is sitting on my back and now I am flying near the water's surface. Then I start to fly higher, which gives me a good view of a well-known archipelago along the Baltic Sea coast. Feeling that I am too high up, I calm myself and slowly descend towards the water and we both dive in. We swim back to the shore and I realize that I should not have flown so high with my daughter.

This dream evokes feelings of freedom, danger, and love. Swimming together symbolizes a new relationship between me and my caring feelings.

Today, various environmental threats are being discussed without any particular focus or context. The fact that excessive nutrients are leaching into the seas, creating eutrophication and anoxic bottom water with dead ecosystems, is well known. However, the severe damage overfishing (Jackson et al. 2001) is causing the marine ecosystem is known only to a small group of scientists. The whole marine ecosystem is strongly degraded by overfishing and populations of large fish have been reduced greatly.

Climate change is widely believed to be causing many problems. A new and growing concern, however, is that increased fossil fuel burning not only influences the temperature of the atmosphere but also increases the acidification of the ocean (Gustafsson et al. 2015; Omstedt et al. 2014; Riebsell and Gattuso 2015). Society is expecting clear and simple communications from scientists, but within the scientific community there is growing understanding of the importance of developing tools to address the combined effects of multiple stressors. This "cocktail" of threats may cause unforeseen and severe damage to ecosystems. If scientists cannot isolate and detect changes, as they are influenced by many factors, or cannot attribute the reasons for these changes, how can they guide society to restore a healthy environment? Working on detection and causal attribution, similar as the global climate change community, inspired me into new research efforts.

References

Gustafsson, E., Omstedt, A., & Gustafsson, B. G. (2015). The air–water CO_2 exchange of a coastal sea: A sensitivity study on factors that influence the absorption and outgassing of CO_2 in the Baltic Sea. *Journal of Geophysical Research: Oceans, 120*(8), 5342–5357.

Jackson, J. B. C., Kirby, M. X., Berger, W. H., Bjorndal, K. A., Botsford, L. W., Bourque, B. J., et al. (2001). Historical overfishing and the recent collapse of coastal ecosystems. *Science, 293*(5530), 629–638.

Omstedt, A., Humborg, C., Pempkowiak, J., Pertillä, M., Rutgersson, A., Schneider, B., et al. (2014). Biogeochemical control of the coupled CO_2–O_2 system of the Baltic Sea: A review of the results of Baltic-C. *Ambio, 43*(1), 49–59.

Riebsell, U., & Gattuso, J.-P. (2015). Lessons learned from ocean acidification research. *Nature Climate Change, 5*(1), 12–14.

Chapter 9
Searching for the Emotional Context

The dreams described in the previous chapters exemplify also a search for identity. I am an identical twin with a brother who was born shortly after me. We were raised according to the doctrine that one should treat twins equally; today, the belief is that parents should support the development of each person's personality. We were treated equally and became dependent on each other. We were dressed in similar clothes and were often confused with each other by our neighbours. We had become The Twins. At the age of fifteen years, we started to diverge and began dressing in clothes that did not look the same, though we often failed in this as we tended to choose very similar styles. Identical twins generally share the same DNA, but already in the womb they start to differ due to random events occurring in all cells. For example, twins do not have the same fingerprints or suffer the same illnesses. My awareness grew of the importance of understanding who I was, but it was not until I began working with my dreams that I could strengthen my own identity and gradually realize that my brother and I were two quite different persons. Dream work provided great support in finding my identity, often activating two equally important parts of my inner mind, one brave and one scared. Literature gave me lots of inspiration, and I was now reading in a new way, looking for myself, my brothers, my family, and my colleagues in the texts I read.

The path to opening up the dream world was through my own vulnerability and feelings, so clearly illustrated by the dream about a scorpion biting my Achilles tendon. To find myself, I had to delve into my childhood memories and early experience. I realized that what was driving me was the human right to be a free person with the corresponding right to take part in society.

During last two decades of my life, I had established a research group at the University of Gothenburg in Sweden to explore several new and interesting subjects, giving me good experience in guiding students. At the same time, I was deeply involved in developing an international Baltic Sea programme, to create an open-minded environment that included scientists from the Baltic States and Russia at the same time as the European Union was developing. We scientists learned a lot from

A. Omstedt, *Connecting Analytical Thinking and Intuition*,
SpringerBriefs in Earth Sciences, DOI 10.1007/978-3-319-27534-5_9

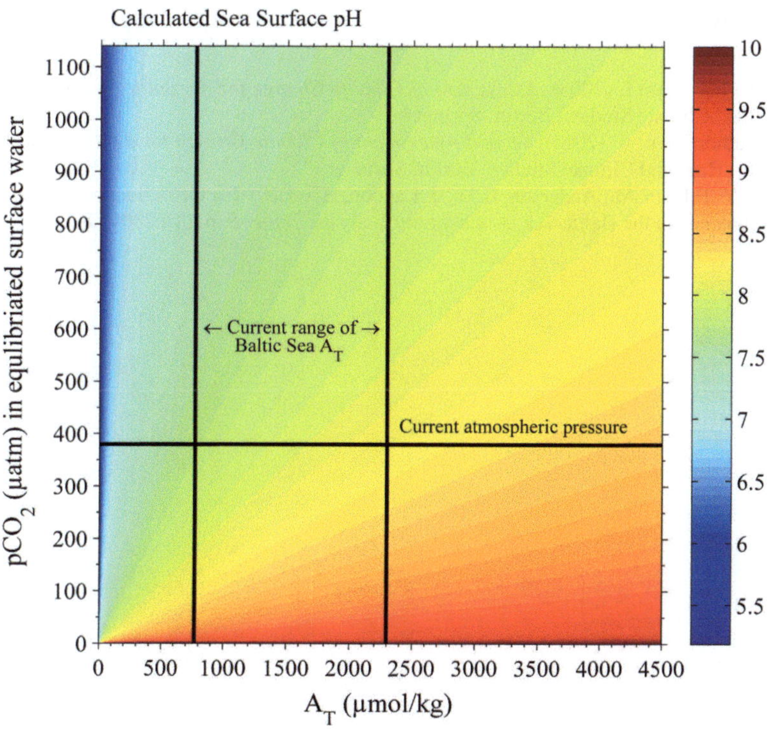

Fig. 9.1 Concern about ocean acidification is increasing, and the phenomenon is expected to occur in coastal seas as well. How this will affect coastal seas and their ecosystems is a new research field. The diagram shows that with increasing concentrations of carbon dioxide and reduced total Alkalinity (A_T), the sea will become more acidic, though much remains unknown about coastal acidification (redrawn from Omstedt et al. 2010)

each other, but the most important lesson was that, by creating a scientific network, we were able to become friends. As we were making great scientific progress, the public management of the seas was increasingly failing. Various threats to the seas, such as overfishing, eutrophication, plastic waste, and climate change, were growing in seriousness, while competition between institutions and universities was increasing (Fig. 9.1). In Henrik Ibsen's 1882 play *An Enemy of the People*, the process of determining scientific truth comes into conflict with society. Many scientists can easily identify with Dr. Stockman who became a hated whistle-blower when he found parasites in the city's main source of income, the water supply for the city's spa. The battle became a bitter one between truth and self-interest. To fight bitter feelings and destructive communication, scientists have now developed strong assessment studies that communicate with society in a new way. I was involved in two such exercises, BACC I and BACC II (BACC Author Team I 2008; BACC Author Team II 2015), and that experience improved my understanding and communication of the strength and the limits of science.

References

BACC Author Team I. (2008). *Assessment of Climate Change for the Baltic Sea Basin.*, (Series: Regional Climate Studies) Berlin: Springer.

BACC Author Team II. (2015). *Second Assessment of Climate Change for the Baltic Sea Basin.*, (Series: Regional Climate Studies) Berlin: Springer.

Omstedt, A., Edman, M., Anderson, L. G., & Laudon, H. (2010). Factors influencing the acid–base (pH) balance in the Baltic Sea: A sensitivity analysis. *Tellus B, 62*(4), 280–295.

Chapter 10
Playback and Meaning

To free myself of feelings of being frozen and trapped by expectations, I started to explore science, literature, and dreams to discover my own feelings and inspiration. I was now seeking the very source of my feelings—to find the Nile River within myself, as it were. Seeking the truth is a journey without a known destination. The philosopher Karl Popper stated that we can know nothing completely; instead, he developed the falsifiability criterion for distinguishing science from non-science. For example, no number of observations of white swans disproves the possibility of black swans—by finding just one black swan, one can easily falsify the statement that all swans are white. By contrast, appreciating the importance of dreams and feelings leads to an "aha" moment. This is not a verifiably true idea but rather a vision or realization that does not follow logical rules. The falsifiability criterion does not apply to dreams, as it is only the dreamer who can interpret their message. In the search for oneself, the memories that surface within dreams and the "aha" moment build our personal identity.

Ullman emphasizes that, in its general features, the unconscious domain is a natural healing system geared to supporting the survival of individuals and the species. This healing function is realized through three special attributes of the unconscious: (1) its capacity to respond creatively to novelty; (2) its concern with our state of connectedness with others; and (3) its capacity to discern truths that escape us while awake (Ullman 1987; 2000).

Let us now review the earlier chapters. The dream about my young son who climbed up onto the deck of a sailboat and fell into the sea illustrated for me the importance of taking responsibility for one's own vision and ideas. One must understand that this is the crucial starting point for any person who wishes to enter the world of science or simply learn more about him or herself. Steering a motor boat from the Baltic Sea into the Kattegat was an image that visualized my new efforts to change course in life and realize that intuition cannot be controlled at will; instead, it is a gift from within, and could be recognized through dreams.

A. Omstedt, *Connecting Analytical Thinking and Intuition*,
SpringerBriefs in Earth Sciences, DOI 10.1007/978-3-319-27534-5_10

In my scientific work, it has become increasingly clear that we are living in the Anthropocene era when human activities have global impacts. As humans, we still lack the tools to handle the challenges of this era and, although much scientific knowledge is available, new ways of thinking and communicating are needed. Scientific initiatives are trying to involve all categories of people in addressing the grand challenges facing the Earth, but we lack the instruments to create and support sustainable change. In the reported dream from Chap. 7, the image of another, but this time an old, sailing ship helped me understand that dreams are not dangerous but illuminating, and that this realization should be shared with others.

My own background and inner feelings have started to find a voice. The dream of walking with my family evoked a feeling of happiness as I reconnected with my inner domain. The beauty of a dragonfly and the attack by the boy illustrated the need to pay full attention. The dream about the foreign children and the cheque for 90,000 pounds reminded me of the need not just to take care of myself but to be aware of how being a twin has influenced my inner space. The inspirational dream about historical barriers and the Internet illustrated a way forward. The wounded horse and the need for a new perspective emphasized the need to stop frustrating myself and change my working conditions. The dream of meeting and drinking water with C.G. Jung illustrated the need to delve deeper into my unconscious mind, supporting my resolve to continue my dream studies. In the dream in which I was singing, I received encouragement to articulate my inspiration and vision. Finally, the dream of flying with my daughter and reconnecting land and sea indicated a new approach free of burdens.

The meaning of dreams lies in our non-logical understanding. Dreams visualize the communication between our outer and inner space, enacting this dramatically in an inner theatre that helps us to find our true self. Dreams can be regarded as the River Nile before the Aswan Dam was built, a rich inner resource that fertilizes our souls. True freedom is—being able to follow our own understanding and intuition.

For me, the scientific journey started with a focus on important physical processes, then building a systemic understanding using mathematical models, teaching, communicating with society, adding chemical and biological processes to the models, and finally exploring the human impact on our natural resources. This journey has given me great satisfaction as well as frustration. Through climate assessment work, I have improved my communication skills and learned the limitations of being a scientist. Most importantly, I felt part of a large research community comprising many open-minded colleagues and friends strongly committed to exploring the unknown.

References

Ullman, M. (1987). Wholeness and dreaming. In B. J. Hiley & F. D. Peat (Eds.), *Quantum implications: Essays in honour of David Bohm* (pp. 386–395). London and New York: Routledge & Kegan Paul.

Ullman, M. (2000). The enduring mysteries of dreams. *Dream Appreciation Newsletter*, *5*(3). Available from http://siivola.org/monte/Dream_Appreciation_Newsletter/2000-3.pdf

Chapter 11
Strengthen Humanity

Many of us are trying to learn from our unconscious dimension, and we all know that we can never fully understand what is hidden there. Many secrets there can be brought to light and understood in various ways, and many methods are available to advance this exploration. Science, religion, literature, art, music, and dreams all illuminate important aspects of being human. Where our conscious and unconscious dimensions meet is the playground of creativity. Of course, hard work is needed and many scientists who explore their subjects in great depth sooner or later need to work on their own personalities. The universities are productive forums for such developmental efforts, the main goals of which are research, education, and outreach. However, due to lack of trust in the research community, society has burdened universities with various administrative measures without focussing on their main goals. At the same time, many researchers are mired in struggling for research funding, self-assertion, and building narrow competing networks. Discussions of university visions are expressed in modern corporate strategic terms and important considerations become fragmented and lost. Scientists learn to sell themselves instead of being of service to human development. Meeting ourselves through others, understanding our own personal vulnerability, and building good memories are the gateways to humanity. Dreams and literature can teach us much about these processes. At the end of *The Brothers Karamazov* by Dostojevskij, one of the brothers gives a speech to boys who have just buried one of their young friends. The speech expresses the importance of having good memories, particularly from an early age; it also states that even if a person has only one good memory, he or she is safe to the end. Often in life we remember only times when things went wrong and when we or others failed. *The Brothers Karamazov*, Dostojevskij's last novel and in recent literature by for example Svetlana Alexievich, underscores the importance of good memories as nuclei for human growth. Dreams interact with memory in a creative way and help us resist fragmentation by focusing on feelings and symbols that are important in our lives.

A. Omstedt, *Connecting Analytical Thinking and Intuition*,
SpringerBriefs in Earth Sciences, DOI 10.1007/978-3-319-27534-5_11

The Brothers Karamazov was written in 1880 when the Little Ice Age was ending in the Baltic Sea region and we were entering a warmer climatic period marked by strong industrialization. By the end of the nineteenth century, people were already noticing changes in fishing in the seas around Sweden. Discussions started and research was organized. Large negative changes in the benthic fauna and ecosystem have occurred since then, and the question of how to attribute the causality of these changes has been discussed vigorously. A major expedition throughout Swedish seas was performed in 1877 to measure seawater temperature and salinity, and comparing those findings with current readings indicates that little has changed to date. Climate change may therefore not be the main reason for present changes in the Baltic Sea. Instead, other anthropogenic factors such as overfishing and eutrophication may have caused most of the damage. Uncontrolled exploitation of our natural resources has exacted a price and there is a growing need to rethink our relationship with the natural world (Fig. 11.1). Alarming messages about future developments may not help, but may instead support the development of an unfeeling and self-destructive society. Reaffirming the contract between science and society is needed (Hooke 2015),

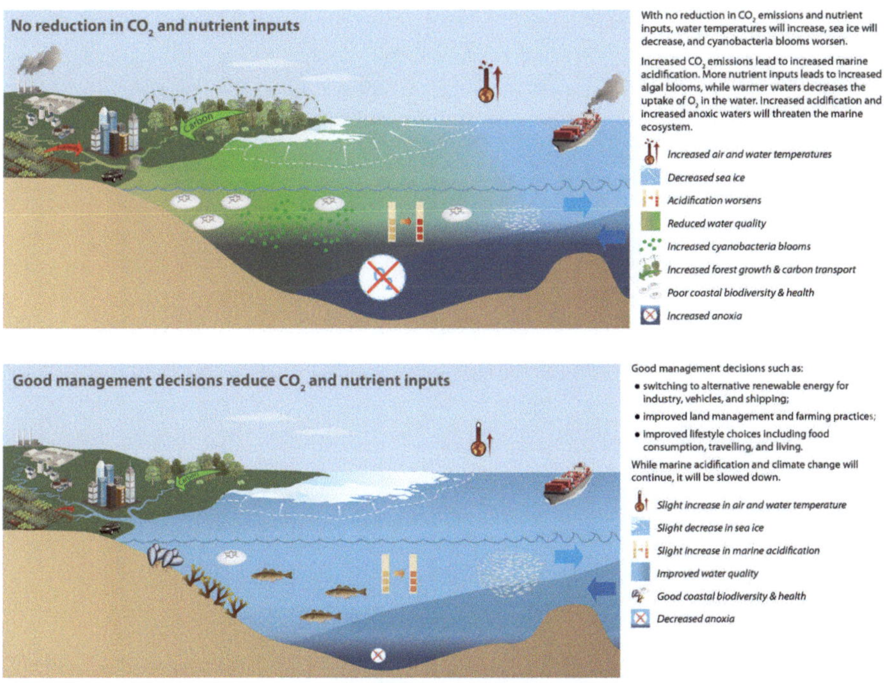

Fig. 11.1 Sketch of possible developments in the Baltic Sea without and with successful management efforts; created by the Integration and Application Network, University of Maryland Center for Environmental Science, USA, with guidance from Omstedt (redrawn from BACC Author Team II 2015)

emphasizing the need not only to seek knowledge and understanding for the benefit of life but for a genuine and loving concern for humanity.

The key task of universities is to help students strengthen their joy in learning, instead of fostering despair about future developments. If universities were able to encourage youth to explore science and their own personalities during their university studies, while remaining fascinated with learning and understanding more about themselves, the future of society would become more secure. In university curricula, including in the natural sciences, teaching that improves students' analytical, empathetic, and self-reflection skills needs to be developed and implemented much better than it is today.

References

BACC Author Team II. (2015). *Second Assessment of Climate Change for the Baltic Sea Basin.*, (Series: Regional Climate Studies) Cham, Heidelberg, New York, Dordrecht, London: Springer International Publishing.

Hooke, W. (2015). Reaffirming the social contract between science and society. *Earth and Space Science News, 96*(6), 12–13.

Chapter 12
Science and Dreams

We are living in a time of accelerating information flows calling for our attention. Every day our brains must accommodate masses of information and deal with the attendant frustration. Do our lives have to become more fragmented or can we find a way to integrate them? Ullman stated that, in principle, there are two ways of knowing the world and our relationship with it: the first is the way of scientific knowledge and the second is the way discerned in our dreams. He also regarded dreaming as an adaptation concerned primarily with the survival of the species and only secondarily with the individual.

At the same time, overselling occurs in the triangle between policy, media, and science. For the credibility of science, scientists are obliged to make their endeavour sustainable. Von Storch listed a number of conditions for the sustainable practice of climate science (von Storch 2012). Obviously, scientists need to admit that scientific knowledge is uncertain and often in need of future revision. There is a need to understand that the public discourse of climate change is driven by two different epistemologies, one of science and its knowledge claims and another of the media. Scientists need to explain that reality is complex and that accepting a complex understanding of climate change dynamics means that unequivocal political conclusions about required or meaningful measures cannot be drawn. Scientific knowledge has clear limitations and in society there needs to be a clear division of responsibility related to science and policy.

How sleep helps consolidate learning and memory remains unknown and has long been discussed and questioned, various theories having been put forward. It seems that a good memory strategy is to transform the information to be remembered into images. The brain remembers images more easily than it does abstractions, so it is unsurprising that so many of us claim to remember people's faces better than their names. Dreams are experienced as sequences of mental images and it is a truism that images often convey much more than texts do. New research provides insights into the strong relationship between our dreams and our memories and demonstrates that the mechanisms that we employ when dreaming are the same as those we use

A. Omstedt, *Connecting Analytical Thinking and Intuition*,
SpringerBriefs in Earth Sciences, DOI 10.1007/978-3-319-27534-5_12

to construct and retrieve memories when awake (Van der Linden 2011). It seems as though dreams help regulate traffic in our brains that connects our experiences with our emotions and memories.

Biological experiments on mice indicate also that sleeping plays a key role in promoting learning-dependent synapse formation and maintenance in selected dendritic branches that contribute to memory storage (Yang et al. 2014). It is obvious that sleeping and dreaming are important for how we cope with our lives; might they not also be important for how we cope with an uncertain future?

Fragmentation can be cured by dream work, giving us a more accurate and integrated perception of life. Dreams convey wholeness and therefore increase our capacity to acknowledge others and nature. How we as humans can shift from a pirate mentality to a caring mentality is the great challenge of our time. Our ability to observe and integrate information from our dreams, however, is still largely unexplored. The interaction between analytical thinking and intuition is not developed in most school and university curricula. New knowledge about our brains and the role of dreams needs to be made available in more education programmes. Training in using both analytical thinking and intuition occurs in dream groups inspired by Ullman, which can serve as an excellent teaching model for improving our creativity capability. This approach has been successfully developed by Stimson to acquaint students and teachers in social work education with their inner reflective resources (Stimson 2009). An important step would be to incorporate the same method into science education, to help students deal with both analytical thinking and intuition as well as develop methods for self-reflection.

References

Stimson, W. R. (2009). Using dreams to train the reflective practitioner: The Ullman dream group in social work education. *Reflective Practice: International and Multidisciplinary Perspectives, 10*(5), 577–587.

Van der Linden, S. (2011, 26 July). The science behind dreaming. *Scientific American*. Available from http://www.scientificamerican.com/article/the-science-behind-dreaming/

Von Storch, H. (2012). Sustainable climate science. In M. Reckermann, K. Brander, B. MacKenzie, & A. Omstedt (Eds.), *Climate Impact on the Baltic Sea: From Science to Policy* (pp. 201–209). Berlin: Springer.

Yang, G., Wan Lai, C. S., Cichon, J., Ma, L., Li, W., & Gan, W.-B. (2014). Sleep promotes branch-specific formation of dendritic spines after learning. *Science, 344*(6188), 1173–1178.

Appendix A
Working with Dreams Using Montague Ullman's Dream Group Process

In the dream group process, described by Ullman (1996) as well as by Siivola (2011), a group of six to eight people work for about 2 hours on one dream, according to a dreamer-centred process. A safe and open-minded environment is created and led by the dream group leader. Nobody other than the dreamer can fully understand his or her dream; accordingly, the method does not aim to explain the dream but rather to teach group members how to appreciate the knowledge conveyed in dreams.

Catch the Dream
At the beginning of a group meeting, one group member presents his or her dream, which has been described in writing before the meeting. Every dream group member writes down the dream as exactly as possible without any additions. The group members then ask clarifying questions to obtain a clear description of the dream.

Finding Words for Feelings and the Meanings of Symbols
The group makes the dream its own and starts to explore the feelings it evokes and later the symbols contained in it. Each member works with the dream and asks: "If this were my dream, I would feel…" and "If this were my dream, the symbols would remind me of …". At this stage, each dream group member projects feelings and metaphors from his or her own experience.

Exploring the Unconscious and Searching for the Emotional Context
In the next step, the dream is returned to the dreamer by letting the dreamer associate freely without interruption. Then a dialogue starts by searching for recent emotional context, for example what happened the day before the dream.

Playback and Meaning
The dream description is read back and new ideas can be explored through open-ended questions. The members summarize and suggest possible connections between the dreamer's dream imagery and waking reality. The dreamer makes a final statement and the group meeting ends. At the next meeting, the dreamer shares any further comments on his/her dream.

A. Omstedt, *Connecting Analytical Thinking and Intuition*,
SpringerBriefs in Earth Sciences, DOI 10.1007/978-3-319-27534-5

Outlook

The method effectively trains group members in analytical thinking and intuition, in empathetic and self-reflection skills. The method can also be applied to a number of other group processes, such as poetry and literature interpretation or inspiring students to solve scientific problems. In the latter case, a group of students starts working by asking whether anyone has a problem he or she would like to share. Letting a student explore his or her own scientific problem in a supportive group, facilitated by a summary from the group leader, can be a very useful step towards solving the problem and helping the student to communicate. It is often useful to ask the student to sketch the problem on the board when describing it.

References

Siivola, M. (2011). *Understanding dreams: The gateway to dreams without dream interpretation*. New York, NY: Cosimo Books.

Ullman, M. (1996). *Appreciating dreams: A group approach*. Thousand Oaks, CA: Sage Publications.

Appendix B
Exercises to Practise Connecting Analytical Thinking and Intuition

Analytical, critical thinking is a key skill emphasized in university training and involves several important aspects, such as identifying main arguments, evaluating these arguments and identifying hidden agendas or flawed arguments. This thinking is slow and limited. Intuition is an insight to know something without any proof or evidence based on your own experience. This is a capability that everyone has and can be trained. It is fast and theory free. Combining analytical thinking and intuition calls for systematic work on both, and for combining them into a relevant story. The following exercises address the different stages of the dream group process. In lieu of dreams, an image chosen from the book or an optional research question is used as the material.

Exercise 1
Select an image from this book. Observe it carefully, and then keep it in your mind as you put the image away. Write a short description of the image. *Hint*: Stay in a quiet room where you can concentrate, sit calmly in a chair, and focus on yourself. Study the image and, if needed, close your eyes before describing it. By doing this, you are training your memory and your ability to observe details. Closing your eyes often better connects you with your own creativity.

Exercise 2
Identify specific parts of the image selected in Exercise 1 along with related symbols based on your memory. Explore the feelings the image evokes in you. Write down your feelings and explore possible associations between them and the related symbols. *Hint*: An image of a burning candle can evoke feelings, such as calmness, warmth, mindfulness, loneliness and happiness. The candle can symbolize guidance, wholeness, hope, peace, etc. Make the image your own by thinking about what it illustrates in your own life and asking what feelings it evokes and what the related symbols suggest. This training in putting words to feelings is an important step in learning how to observe your own reactions. In addition, you are training yourself in empathetic thinking, i.e. understanding the implications and connotations beneath the object's surface appearance. A candle is a candle but can be understood in many different ways.

A. Omstedt, *Connecting Analytical Thinking and Intuition*,
SpringerBriefs in Earth Sciences, DOI 10.1007/978-3-319-27534-5

Exercise 3
Explore your own life situation by writing down your feelings about these exercises, what happened the hours before you come in contact with this exercise, the day before, a week ago? *Hint*: You probably did not choose the image by chance but for a reason. In all life situations, we carry with us many feelings from the past that influence how we react. Here you are training yourself to observe your own feelings and how these may influence your decisions.

Exercise 4
Reread what you have written and add any new ideas about the image and why you decided to choose it. *Hint*: Read slowly and reflect on what feelings and awareness the text, coupled with your life situation, is evoking in you. Here you are training yourself to deepen your analytical thinking and strengthen your intuition through reflection. Often you discern more than expected when rereading, indicating the value of deep reflection, respecting one's feelings, and allowing sufficient time for thinking.

Exercise 5
Write a summary based on the image and your reflections on it. Create an evocative title for the text. *Hint*: Write the summary in a somewhat dramatic way and let the title summarize the vision created in you by the image. Here you learn to put your feelings, life situation, and intuition together in an effective way. This exercise trains your ability to integrate information in a coherent way, the way a chairperson does who has been listening intently to all the arguments expressed in a meeting. You are training yourself to assimilate not just what has been explicitly said but also what has been implied in a new and creative way.

Exercise 6
Select a scientific issue that you would like to understand in greater depth. Sketch a diagram of the problem you would like to address. Examine your feelings and the reasons why you have chosen this problem. Consider your life situation in terms of the skills that you need to improve, the time you have available, and the guidance you may get from your surroundings and from the library. Formulate a short title encapsulating your scientific problem, phrased, for example as a question. Write a summary including the diagram of the problem. In this exercise, consult helpful and reliable teachers/scientists. *Hint*: A research group can be very helpful when you are addressing a new problem, as you can ask its members to imagine how they would solve the problem if it were theirs. When exploring the scientific issue, record new findings and feelings by expanding your summary into an article structure. Organize your thinking by, for example preparing a PowerPoint presentation. A major step towards success is when you can make a research problem your own. By "pushing" gently on the problem and treating it as your own research project, you will deepen your understanding in your problem. You should, however, be careful not to hurt yourself and not just to work in an analytical way. Instead, you should reserve time to enjoy life and play with new ideas, feelings, and colleagues. Be open to ideas and feelings and push your thinking to its limits.

Appendix C
Solutions to Exercises

No single solutions can be given to the above exercises, as it is only the reader who can fully understand the solutions he or she devises. Below I will illustrate possible solutions by relating them to my own experience.

Exercise 1
Select an image from this book. Keep it in your mind as you put the image away. Write a short description of the image.

Solution:
I selected Photo (in Appendix B), which depicts a room with an open door to the right. The room is dark, and light is streaming into it through the door. The door is white and inside the room part of the floor is visible. The visible floor appears old and displays a pattern of square segments or tiles.

Exercise 2
Identify specific parts of the image, as well as related symbols, based on your memory of it. Explore the feelings the image evokes in you. Write down your feelings and explore possible associations of the various related symbols.

Solution:
Dark rooms give me feelings, such as calmness, peace, relaxation and loneliness. The open door awakens feelings of hope and curiosity while the floor gives me feelings of stability and trust. The image includes a dark room, an open door, a floor and light coming in from right. Inside me, the dark room symbolizes a hidden place where dreams and feelings exist. The door illustrates a connection to something outside me that now it is open. The light symbolizes opportunities and new challenges. The floor confidence that I can take advantage of these new opportunities.

Exercise 3
Explore your own life situation by writing down your feelings about these exercises, what happened hours before you come in contact with this exercise, the day before, a week ago?

A. Omstedt, *Connecting Analytical Thinking and Intuition*,
SpringerBriefs in Earth Sciences, DOI 10.1007/978-3-319-27534-5

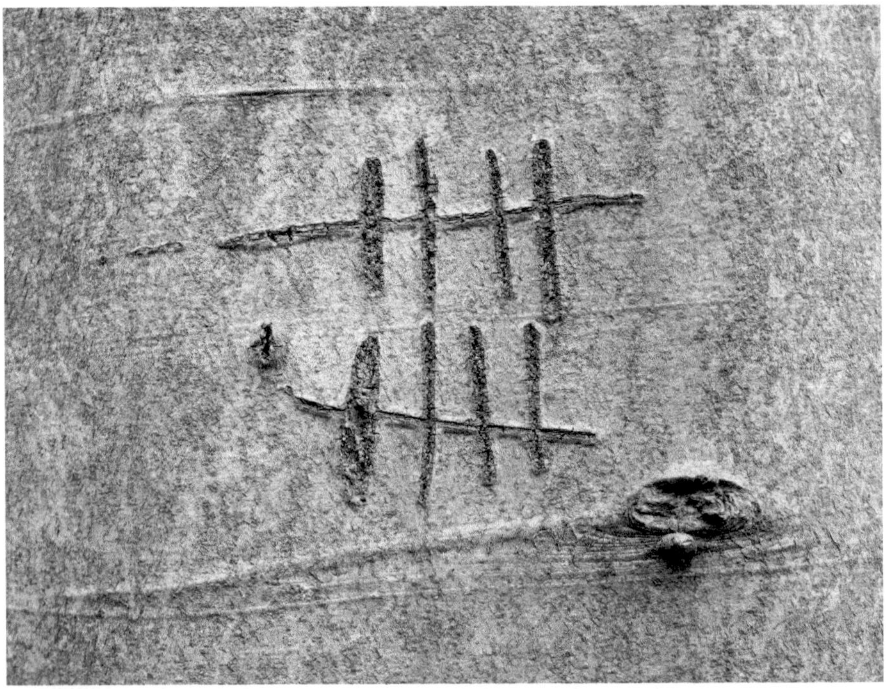

Solution:
The day before I started this exercise, I received the photos for this book from Hillevi Nagel, which was very exciting and stressful. Before choosing the images, I worked a whole evening collecting the best photos for the book. I was pleased to be doing this and concentrated well. Over previous weeks of work on this book, I had been feeling inspiration but also frustration at not knowing when and where it would be published. I had received positive responses from two friends who had read parts of the book, which gave me confidence in the project.

Exercise 4
Reread what you have written and add any new ideas about the image and why you decided to choose it.

Solution:
Rereading makes me realize that writing is very lonely work. No people are present in the image, which illustrates that I am alone in this activity. The open door is promising, however, and what does it mean that the light is coming in? Perhaps the dark portion of the room represents intuition, located in my unconscious mind, while the door and light represent my conscious intellectual and language abilities? If so, then the connection is the floor, which seems to be from an older time, representing a path that others have taken before me.

Exercise 5
Write a summary based on the image and your reflections on it. Create an evocative title for the text.

Solution:
Title: Time to connect to others.
The room is dark and empty without any people, but light is shining into the room through a white door, revealing patterns, apparently of old tiles, in the floor. The dark room gives me the feeling that I can be calm and rely on what is coming to me through the open door. The dark room symbolizes a hidden place inside me where dreams and feelings exist. The door illustrates a connection to something outside me, a connection that now it is strong. The light symbolizes possibilities and new challenges, as the lonely work of writing this book will soon end. The door is open between my unconscious mind and my communication ability, and several others have walked this way before me. This is an experience I share with many other authors and scientists.

Exercise 6
Select a scientific issue that you would like to understand in greater depth. Sketch a diagram of the problem you would like to address. Examine your feelings and the reasons why you have chosen this problem. Consider your life situation in terms of the skills that you need to improve, the time you have available and the guidance you may get from your surroundings and from the library. Formulate a short title encapsulating your scientific problem, phrased, for example as a question. Write a

summary including the diagram of the problem. In this exercise, consult helpful and reliable teachers/scientists.

Solution:
I have done such exercises many times and found them very rewarding. Now it is up to the reader to undertake this exercise.

MIX
Papier aus verantwortungsvollen Quellen
Paper from responsible sources
FSC® C105338

If you have any concerns about our products,
you can contact us on
ProductSafety@springernature.com

In case Publisher is established outside the EU,
the EU authorized representative is:
Springer Nature Customer Service Center GmbH
Europaplatz 3, 69115 Heidelberg, Germany

Printed by Libri Plureos GmbH
in Hamburg, Germany